YOUR KNOWLEDGE HAS VALUE

- We will publish your bachelor's and master's thesis, essays and papers

- Your own eBook and book - sold worldwide in all relevant shops

- Earn money with each sale

Upload your text at www.GRIN.com and publish for free

Vinay Divakar

Mechanical Aspects in Electronics Systems Design

GRIN Verlag

Bibliografische Information der Deutschen Nationalbibliothek:

Die Deutsche Bibliothek verzeichnet diese Publikation in der Deutschen National-
bibliografie; detaillierte bibliografische Daten sind im Internet über http://dnb.d-
nb.de/ abrufbar.

Imprint:

Copyright © 2013 GRIN Verlag GmbH
Druck und Bindung: Books on Demand GmbH, Norderstedt Germany
ISBN: 978-3-656-73090-3

This book at GRIN:

http://www.grin.com/en/e-book/279301/mechanical-aspects-in-electronics-systems-
design

GRIN - Your knowledge has value

Der GRIN Verlag publiziert seit 1998 wissenschaftliche Arbeiten von Studenten, Hochschullehrern und anderen Akademikern als eBook und gedrucktes Buch. Die Verlagswebsite www.grin.com ist die ideale Plattform zur Veröffentlichung von Hausarbeiten, Abschlussarbeiten, wissenschaftlichen Aufsätzen, Dissertationen und Fachbüchern.

GRIN - Your knowledge has value

Visit us on the internet:

http://www.grin.com/

http://www.facebook.com/grincom

http://www.twitter.com/grin_com

ASSIGNMENT

Module Code	ESE2515
Module Name	Mechanical Aspects in Electronic System Design
Course	M.Sc in Electronic System Design Engg
Department	Electrical and Electronics Engineering

Name of the Student	Vinay Divakar
Batch	Full-Time 2012

M.S.Ramaiah School of Advanced Studies
Postgraduate Engineering and Management Programmes(PEMP)
#470-P Peenya Industrial Area, 4th Phase, Peenya, Bengaluru-560 058
Tel; 080 4906 5555, website: www.msrsas.org

POSTGRADUATE ENGINEERING AND MANAGEMENT PROGRAMME – (PEMP)

The demand for opto-electronics is increasing as we are nearing the future due to the need for high data rate, high bandwidth, lossless transmission and low electromagnetic interference sensitivity. In **chapter 1**, the present research carried on the mature laser technology i.e. GaAs, in order to improve its efficiency. The packaging principle used for receivers can be applied for the packaging of the laser driver circuit and the laser source in a single module. The concept of FRACAS (Failure Reporting, Analysis and Corrective Action System) has been described and failure analysis technique for Electrical overstress (EOS) is described. An industrial approach to calculating the reliability of a system with some known data is described. Some challenges with respect to packaging has been discussed in detail and some methods to overcome challenges such as lattice mismatch has been described.

Every electronic components or electronic systems have certain specifications based on which it is developed and all components have datasheets of their own. The datasheets consists complete details related to the product such as product design specifications, packaging type, power ratings, dimensions etc. Any components can be selected for a particular application by referring their datasheets. In **chapter 2,** a datasheet for a DC power supply has been developed covering most of the important details that may be needed for designing and modeling using software tools. The schematic of the power supply is developed and practical tests are performed on the power supply, which has been described in details with the test results. The power supply has four interfaces and the functionality and usability of these interfaces has been shown and described in detail.

Before the large scale manufacturing and production of any product, it is necessary to conduct two basic tests i.e. Thermal analysis and vibration tests for any given product. These tests help us to get an insight to the reliability of the product. In **chapter 3**, the power supply is modeled using the software tool ICEPACK v13, using which thermal analysis is performed on the critical components and the temperature variation curves along with the simulation results has been discussed. The method of casing used for the power supply for modeling and the types of conventions i.e. natural and forced convention systems has been compared and discussed. An experimental set up used for performing vibration testing on the power supply has been demonstrated and described in detail.

Contents

List of Tables

List of Symbols and Acronyms

A	Amperes
AC	Alternating Current
AlSiC	Aluminum Silicon Carbide
CTE	Coefficient of Thermal Expansion
DC	Direct Current
DRC	Design Rule Check
EI	Electrical Interconnect
EMI	Electromagnetic Interference
EMC	Electromagnetic Compatibility
FMEA	Failure Mode Effect Analysis
FTA	Failure Tree Analysis
GHz	Giga Hertz
HAZOP	Hazards and Operability Analysis
IC	Integrated Circuit
MCM	Multi Chip Module
ITRS	International Roadmap of Semiconductors
MOEMS	Micro-Optical Electro Mechanical Systems
OI	Optical Interconnect
OE	Optoelectronics
OEIC	Optoelectronic Integrated Circuits
PCB	Printed Circuit Board
R&D	Research and Development
TEC	Thermoelectric Cooler
VCSEL	Vertical Cavity Surface Emitting Laser
UV	Ultra-Violet
WLP	Wafer Level Packaging
Si	Silicon
μ	Micro
mA	milli Amperes
V	Voltage
°C	Degree Centigrade
Hz	Hertz

1. Opto-electronics packaging and Failure Analysis Methodologies

1.1 Recommendations for Opto-electronics packaging design

Packaging is nothing but a sequence of process steps involving connecting, protecting and manufacturing of the devices. The widespread commercial utilization of semiconductor lasers have now included the opportunities to make use array of diode lasers as monolithic components on the Silicon Integrated Circuits (IC's). Presently the laser manufactures are optimizing the design and process to maximize the laser performance for InGaAsP wafers producing around 10,000 lasers per square inch of the wafer for a compact opto-electronics module packaging. Receivers have been developed which consists of an array of photodiodes along with pre-amplifiers in one module mounted on Silicon substrate in one single package, where one side of the photodiode's receive photons and convert it into the desired current or voltage, while on the other side it has electrical contacts in order to trigger the desired component or IC, therefore the same packaging principle can be applied for the array of transmitting lasers that improves manufacturability reducing cost, which has been demonstrated in [Mino F. Dautartas 2002 IEEE]. Another important aspect of Opto-electronic packaging is the wire bonding. Many Opto-electronics packaging is designed as "butterfly" shape which deals with both electrical and optical signals. The Electrical interconnections are from the cantilevel leads to the pads of the die mounted inside the package, where the height difference between the cantilevel leads and the die pads are large, therefore it requires the wirebonds to have the capability of deeper access and typically wire bonding in opto-electronic packaging has the first bond on the cantilevel leads and the second bond on the die pads due to the bond pads on the cantilevel leads being very close to the package walls, therefore to avoid interference of the wedge with the package walls the first bond is normally placed on the leads, and there are three wirebond technologies that are used for opto-electronics packaging i.e. thermo-compression bonding, thermo-sonic bonding and ultrasonic bonding as illustrated in [jianbio pan 2004]. The ability to faithfully reproduce designed features on a wafer is important to the success of optical devices. Critical Dimension (CD) control and smooth device surfaces are two of the most important parameters that must be controlled during processing. Resolving CDs of approximately 350 nm is hardly a challenge for modem optical lithography processes, whose minimum features have already been demonstrated below 45 nm. Resolution, however, is not the only concern when designing planar optical circuits. The need for better resolution demands flat surfaces on which to pattern features. This requirement comes about as a result of the usable Depth of Focus (DOF) at the image plane of standard projection optics lithography equipment; 90 nm process technology has a usable DOF < 300 nm. As a result, wafer topography must be kept below the DOF value to successfully resolve minimum feature sizes.

1.2 Recommendations for Failure Analysis Technique

Every product or a system has modes of failure. It is important and necessary to understand why a failure has occurred or why a component has failed in order to rectify the failure and this process is known as FRACAS (Failure Reporting, Analysis and Corrective Action System) [Walter Willing, Jonathan 2012].A common failure fault that is encountered during electrical testing is the components damage in a system that is caused due to electrical overstress. Electrical Over-Stress (EOS) is a term used to describe the thermal damage that may occur when an electronic device is subjected to a current or voltage that is beyond the specification limits of the device. During the testing the Voltage v/s Current (VI) characteristics curve is drawn or traced for the input to each components present in the system in order to identify if there's any electrical input overstress have occurred. The identified failed parts VI characteristics can be compared to the VI characteristics of the good part and if there's any deviation from the desired response, then the test results are noted and recorded for later examination of that failed part in order to rectify the fault. In industries, all systems that are developed have to undergo environmental tests, mechanical tests and electrical tests in order to detect faults, analyze and rectify them, these test procedures have been illustrated in [MIL-STD-883F 1996]. Another common technique apart from weibull distribution and exponential distribution model is the reliability prediction of a system, where the failure rate of the components and the reliability of the system is calculated. Let's Assume that 600 parts where stressed at 150°C ambient for 3000 hours with one failure at 2000 hours for a photo resist flaw (0.7eV) and one failure at 3000 hours for an oxide defect (0.3eV); the internal temperature rise (Tj) of the part is 20°C and the product was tested at 1000, 2000 and 3000 hours. Then to find the FIT rate for the process with M=6.3 (chi factor distribution for DOF = 2r+2 for r=2) at 55°C, it is necessary to calculate the Acceleration factors AF_1 and AF_2 due to the faults and then the Total Device Hours is calculated followed by the calculation of systems Failure rate in FIT and then the life time of the product or system as illustrated in detail [William J. Vigrass 1997]. These predictions are used to evaluate design feasibility, compare design alternatives, identify potential failure areas, trade-off system design factors, and track reliability improvement.

1.3 Merits and Demerits of Packaging design and failure analysis techniques

In photonics, one of the limitations is the lack of monolithic long lived laser on silicon, therefore from several years the main effort has been focused on the growth GaAs on silicon, with associated problems related 4% lattice mismatch the two materials i.e. GaAs and Silicon. GaAs or InP compounds exhibit a larger lattice constant than that of Si. The only exception is GaP, which has an indirect bandgap and is not suitable as a laser material. Therefore one of the major roadlocks for further development of optoelectronics with respect to materials is the lattice mismatch between two materials or between the films and the available substrates. Mismatched lattice leads to a high defect density (in the 10^8 - 10^{10} level). In spite of the intense efforts made to address this problem, it may not be feasible to reduce the defect density by several orders of magnitude in the conventional approach, until lattice matched substrates are available. Another issue that can be encountered during packaging for optoelectronics is due to the growing density of functionalities and complexity of interfacing interdisciplinary functions that needs to be integrated in a single system i.e. mechanical, electronics and optics. Proper packaging considering thermal issues, vibrations and shock analysis will lead to a mechanical structure for opto-electronics that is highly reliable and whose survival rate will be longer, thus making the system more reliable. Integration problems related to chemical contamination are likely to pose practical problems due to physical limitations caused due to issues like corrosion. Most conflicts can be addressed by incorporating extra de-contamination steps or adding dedicated tools for the problematic operations, but these solutions can lead to increase in cost. Problems due to chemical contamination have been brought under control in standard silicon processing, but would need to be checked on if the process demands the inclusion of elements and compounds not used in standard processing, but even after its introduction over four years ago, the semiconductor industry is still trying to rectify issues associated with copper (Cu) contamination. There are still no universally acceptable levels of Cu contamination, and many questions still remain about how far beyond the fabrication tools and chambers one must de-contaminate, therefore similar logistical problems may be waiting for the monolithic integration of optical devices with electronics and its packaging.

1.4 Essential modifications for the improvement and realization of the product

The system design can be improved by making an effort to overcome the problem of lattice mismatch between the GaAs and the Silicon substrate by making use of the available alloys. Optimization of electronic or optical devices require the capability of forming alloys and heterogeneous structures that can provide improvement in lattice matching. Therefore alloys were developed and examined by making use of Nitride compounds such as GaAIN, GaInN. The combination of a mature technology such as GaAs along with GaN to form GaAsN will allow to produce light emitting alloys with a wider and better range of lattice constant[J. Salzman and I. Samid 1996]. Thermal budget limits are critical when integrating different devices into the same process. Failing to account for thermal budgets could result in the interdiffision of dopant species, weakening of metal layers, and the introduction of stress due to differing coefficients of thermal expansion. Thermal budgets for both aluminum and copper back-end processes with oxide-based dielectrics are limited by the metal and Interlayer Dielectric (ILD). Although some low-k dielectrics (k < 3.9) are being used (e.g. Carbon-Doped Oxide (CDO)), the thermal stresses and the weaker mechanical properties of CDO will not be able to tolerate temperatures in excess of 450°C; low-k solutions such as Spin-on- Dielectrics will drive the thermal budgets even lower. In the case of electro-optical (EO) polymers, the thermal budget can decrease dramatically, since typical glass transition temperatures are in the range of 150°C - 250°C. For all practical purposes, these polymers can only be integrated as the last steps of the process. The 12 channel 3.125 Gb/s VCSEL laser driver IC has been designed using SiGe BICMOS process technology. This miniature laser driver IC can be used for development of opto-electronic systems. The driver IC is integrated in a module with an array of 12 VCSEL laser sources. This driver IC's results has been measured and evaluated[Allan Armstrong, Scott killmeyer IEEE 2001].

3.5 Conclusion

The laser manufactures are doing optimizations for the laser technology and its performance. Principle packaging for receivers can be applied for packaging of laser driver circuit with laser sources. An important aspect of optoelectronics packaging is wire bond. A basic methodology for failure analysis is the FRACAS. An industrial approach for calculating the reliability of the system is done using the formulas of acceleration factor, failure rate and concept of MTTF. EOS can be detected by plotting VI curve for the input of each component and comparing it to a good known reference curves. An important limitation of the laser source packaging for optoelectronics as the lattice mismatch constant for which alloys are developed using nitrate compounds for better lattice matching of the materials. A laser driver IC is integrated with array of 12 VSCEL laser sources using the BiCMOS process technology which can be packaged into one single miniature module that can be used for the development of optoelectronic system.

2. Datasheet development and testing of an electronic product

2.1 Development of data sheet for Linear Power Supply

The datasheet was developed for the DC linear power supply available in the hardware lab of MSRSAS. Some general details regarding the power supply has been described in the table

General parameters	General Description
Size and weight	Heatsinks for high power linear regulators add size and weight. Transformers, if used, are large due to low operating frequency (mains power frequency is at 50 or 60 Hz); otherwise can be compact due to low component count.
Output voltage	With transformer used, any voltages available; if transformer less, not exceeding input. If unregulated, voltage varies significantly with load.
Complexity	Unregulated may be simply a diode and capacitor; regulated has a voltage-regulating circuit and a noise-filtering capacitor; usually a simpler circuit (and simpler feedback loop stability criteria) than switched-mode circuits.
Power Factor	Low for a regulated supply because current is drawn from the mains at the peaks of the voltage sinusoid, unless a choke-input or resistor-input circuit follows the rectifier (now rare).
Efficiency, heat and power dissipation	If regulated: efficiency largely depends on voltage difference between input and output; output voltage is regulated by dissipating excess power as heat resulting in a typical efficiency of 30–40%. If unregulated, transformer iron and copper losses may be the only significant sources of inefficiency.
Radio frequency interference	Mild high-frequency interference may be generated by AC rectifier diodes under heavy current loading, while most other supply types produce no high-frequency interference. Some mains hum induction into unshielded cables, problematical for low-signal audio.
Risk of equipment damage	Very low, unless a short occurs between the primary and secondary windings or the regulator fails by shorting internal

Table 2. 1 General Description

2.1.1 Product Design Specifications

PARAMETER	CONDITION/DESCRIPTION
Input Voltage	AC input of 230 Volts
Input Frequency	AC input 60Hz
Output voltage	
Line Regulation	+/-1%
Output Voltage range	+/-10%
Leakage current	50uA
Operating Temperature range	At 100% load, it is 0-70
Storage Temperature range	-40 to 85
Transient Response	Recovery time to with 1% of initial set point
Performance features	• Dual supply of 5 Volts and 12 Volts

	<ul><li>Minimum output Voltage ripple</li><li>Low leakage current</li><li>Excellent heat dissipation by making use of heatsinks</li><li>Convention cooling (Natural convection)</li><li>Low EMI</li><li>Output voltage regulation</li></ul>
Efficiency	<ul><li>Low Cost</li><li>2 years warranty</li><li>Low maintainance</li><li>Can withstand shocks</li><li>Overload protection</li><li>Light weight</li><li>100% burn in and safe</li><li>Removable top lid for better heat transfer or convection</li></ul>

Table 2. 2 DC Power supply specifications

COMPONENTS	FEATURES/SPECIFICATION
IC 7805 Voltage Regulator	<ul><li>Output voltage of 5 Volts</li><li>Thermal overload protection</li><li>Output current upto 1A</li><li>Short circuit protection</li></ul>
Capacitors (22uF to 6600uF)	<ul><li>Low impedance</li><li>Operating Temperature range -40 to 105</li></ul>
Capacitors (5pF to 2200pF)	<ul><li>Operating temperature range -30 to 85</li><li>Small size and long Leads</li></ul>
IC LM7812 Voltage Regulator	<ul><li>Output voltage of 12 Volts</li><li>Max output current of 1.5A</li><li>Thermal overload and short circuit protection</li></ul>
Transformer	<ul><li>Input Voltage 230V/60Hz</li><li>Isolated Primary and secondary windings</li></ul>
Diode IN4007	$V_F = 0.68$ Volts

Table 2. 3 Component specifications

2.1.2 Brief Introduction to the DC power supply

The Linear DC power supply has output voltages of 5 Volts and 12 Volts, which are regulated by using the highly reliable voltage regulators i.e. IC L7805 and IC L7812, thus generating a highly regulated DC power and additional LC filters are used for generating a smoothened DC power output. It consists of a removable a removable top lid or cover allowing for better heat dissipation.

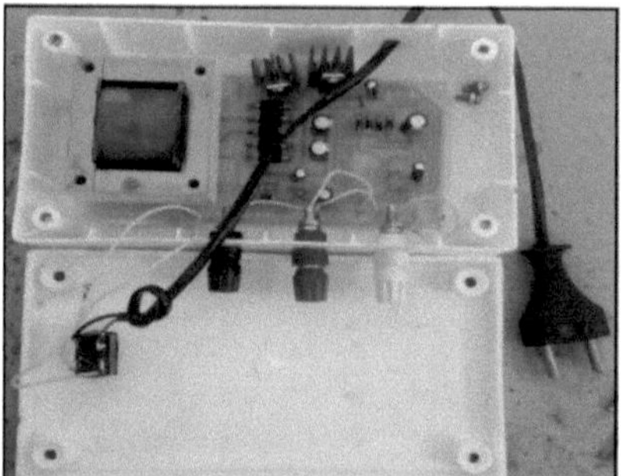

Figure 2. DC power supply

Additional heat sinks have been used by the regulators as shown in the Figure 2, which provides overload protection preventing the device from thermal damage. This power supply is most appropriate in use by low voltage or power applications. It is safe and has Low EMI, thus approved to domestic regulatory standards. It has three terminals i.e. ground, 5 Volts and 12 Volts outputs. Apart from these, the power supply is small and of light weight, hence it is easily portable and the power supply is designed such that its simplicity allows for easy accessibility to the on board Power supply components for maintenance and servicing.

2.1.3 Packaging Details

COMPONENTS	DETAILS
Regulator IC's i.e. IC7805 and IC L7812	• T0-220 style of electronic component packaging i.e. Through hole technology type package • Package top has metal tap with hole needed for heat sink mounting
Capacitors (3300uF)	• Disc ceramic capacitors • Conformably coated • Long and Radial leads (1.0 inch Min)
Capacitors (100nF or 0.1uF)	• Package type A and series FM • Through hole technology • Lead type is Radial
Transformer	• Three Phase oil immersed transformer in hemetric design • Surface mount technology • Natural cooling with copper winding inside a base frame
Diode	
Electronic board	Material - FR 4 Class A

Overall system (Power Supply)	Box type plastic encasing with removable top

Table 2. 4 Packaging Details

2.1.4 Dimensions of the system

Length x Breadth x Height (LxBxH)

Board and Components	Dimensions in [mm] (LxBxH) and Diameter (D)
Electronic board	140x75
Regulator IC L7812	19.1x10.36x4.9
Regulator IC 7805	18.95x9.9x4.5
Capacitor (3,300uF)	L : 20 Body, D : 12.5 Lead, D : 0.8 and L : 17
Capacitor (100nF or 0.1uF)	L : 14 Body, D : 12 Lead, D : 0.7 and L : 25.4
Transformer (5VA)	Body : 37.7x44.5x32.5 Core : 30x38x24
Diode IN4007	L : 5.21 D : 2.72

Table 2. 5 PCB and Components Dimensions

2.1.5 Min/Max current, voltage and power ratings

PARAMETERS	Min	Nom	Max	Unit
Input Voltage	195	230	265	Volts (Vac)
Input Frequency	47	60	63	Hertz (Hz)
Line Regulation	-1.0		+1.0	%
Output Voltage Regulation	-10		+10	%
Leakage current		20	50	uA (micro Amps)
Output Voltage Ripple and Noise			3.0mV PK-PK and 0.2% PK-PK	Milli Volts (mV)

Table 2. 6 Power ratings of the Power supply

2.2 Schematic Development

The Figure 2.1, shows the schematic of the DC power supply. This schematic was developed using the software tool Design Entry by Cadence v16. The main components present in the schematic are transformer, two voltage regulators, H-Bridge using diodes and filters. The output voltage regulators are IC 7805 and IC LM7812. The transformer used is of 230V/60Hz and an LC filter has been used to filter the output of LM 7805.

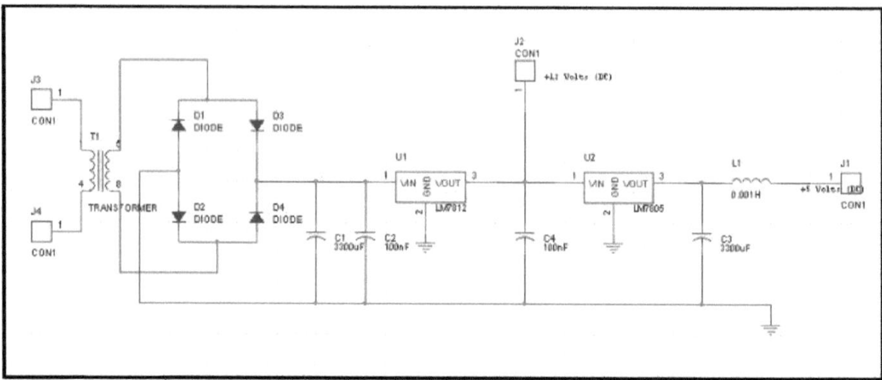

Figure 2. 1 DC power Supply Schematic

The Transformer (T1) primary windings is connected to the 230V,60Hz AC main and the secondary windings are connected to the Diode H-Bridge. The transformer steps down the AC 230 Volts to AC 15 Volts (V) which is rectified by the Full Bridge Rectifier (H-Bridge) i.e. conduction during both –ve and +ve half cycles), therefore a Pulsating DC is generated. These Filter capacitors C1, C2 and C4 are used to ground AC components, blocking only the DC components to pass thus reducing ripples. Due to some voltage drops across these capacitors, the input (Vin) to the regulator IC LM7812 will be approximately 13.5V, this 13.5 V is regulated to 12Volts output (Vout) by the LM7812. Similarly the IC LM7805 is used to regulate the voltage to a +5 V output. An LC filter circuit has been used for further filter of the AC components in order to obtain a smoothened DC output Voltage, the inductor is used to block any incoming AC components, while allowing only the DC components to pass. The values of the capacitors and inductors were designed for an input of 230 Volts and output voltages of 12V and 5V.

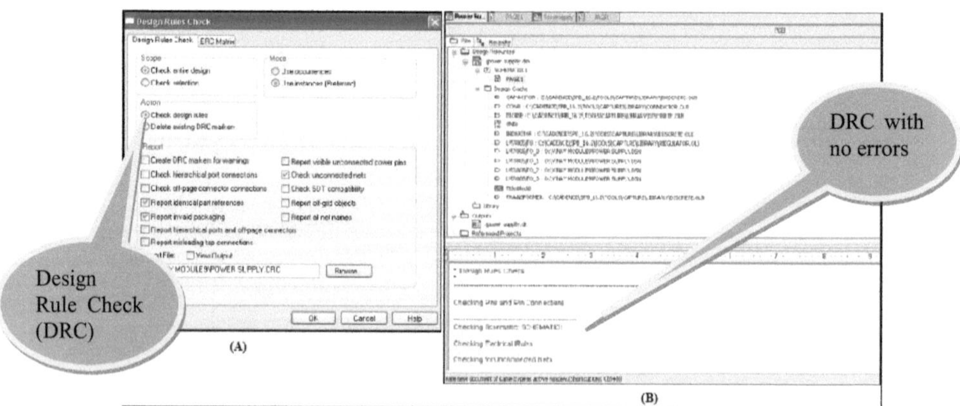

Figure 2. 2 Design Rule Check (DRC)

Once the Schematic is developed, it is necessary to check and know if the developed schematic is correct, therefore a Design Rule Check (DRC) is performed on the schematic. The DRC icon is clicked to run the process as shown in Figure 2.2 A, and the Figure 2.2 B, shows the completion of the DRC of the schematic with no errors, hence schematic can be used for further designing and for development of PCB for the power supply.

2.3 Interfaces Available

The interfaces developed for the power supply helps the user to connect low power devices to the power supply in order to supply power to that electronic device i.e. Used to power up electronic devices. The Figure 2.3, shows the power supply unit with its interfaces and hardware. The power supply has a total of five interfaces i.e. three terminals (GND, 12 Vout and 5 Vout), a Double pole- Double throw (DPDT) switch and an plug to connect to the AC mains to draw 230 V input.

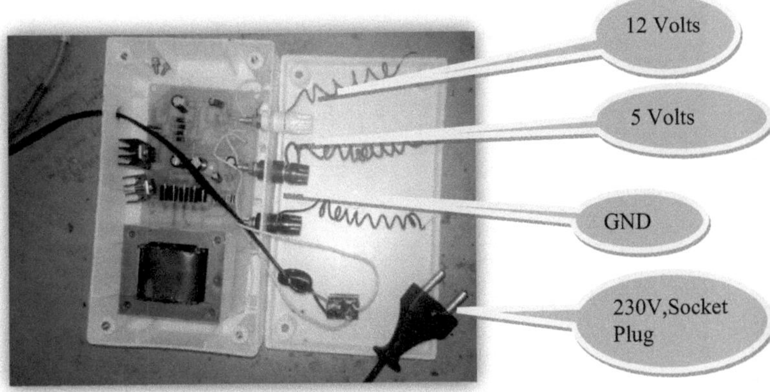

Figure 2. 3 Power Supply Interfaces

The Figure 2.3, shows the power supply with three terminals among which the black connector is the Ground (GND), yellow connector is the 12 Vout from IC L7812 and the red connector is the 5 Vouts from IC L7805. The wires connected to these three terminals are used for interfacing to devices or circuits to supply to power to the devices.

Figure 2. 4 Switch Interface

The Figure 2.4, shows the bottom and the top part of the DPDT switch. In the bottom Part, two terminals are connected to the two pin plug that is used to interface or plug to the AC mains and the other two terminals are connected to the primary windings of the transformer. The top part of the switch is the user interface for user to control the switching ON or OFF of the power supply.

2.4 Components details with Cost and Availability

The Table 2.8, shows the cost and the specifications of the different components being used in this design of the DC power supply. The cost of the components slightly varies from place to place depending on the quantity of each component being purchased. Thus the price range is showed in the table, for those components whose price per unit is not mentioned.

Components	Cost (Rupee) and Qty	Details	Availibility
Electrical Transformer	110/-	230 V to 15 V Step-down transformer	YES
L7812 Voltage Regulator	10.50/-per unit (piece)	Output Voltage : 12V Maximum current : 1.5A Maximum power dissipation : 15W	YES
Capacitor 3300 uF	14/- per peiece 5 or more 13.5/- 10 or more 13/-	85°C Max Temperature Radial electrolytic capacitor	YES
Capacitor 100nF or 0.1uF	1.5/- per Piece	0.1uF ceramic disc shaped capacitor with radial leads	
Inductor	7.5/- per peice 5 or more 7.25/- 10 or more 7/-	1mH or 0.001H	YES
L7805 Voltage Regulator	26/- per unit (Piece)	Maximum Output Voltage: 5Volts Output Voltage fixed: 5Volts Drop Voltage: 2Volts Minimum input Voltage: 8Volts Maximum input Voltage: 20Volts	YES

Table 2. 7 Components Details and Cost

The components mentioned in the Table 2.8, are all available at electronic shops in Bangalore. Some common known shops that sell all these components are; Leed Electronics Industry Inc, JP nagar-560078 and NSK Electronics, opposite to Vishal electronics, SP road-560002. All these components are readily available or can be per-ordered in these respective shops.

2.5 Test set up and results

Tests are performed on the DC power supply available in the MSRSAS, hardware lab in order to check and analyse the response of the system or power supply. The test analysis was mainly based on checking the output voltages and load currents using a 12 Volts DC motor. The Figure 2.5 A, shows the testing of the output voltage from the 5 Volts connector that is displayed on the multimeter i.e. 5.51 Volts.

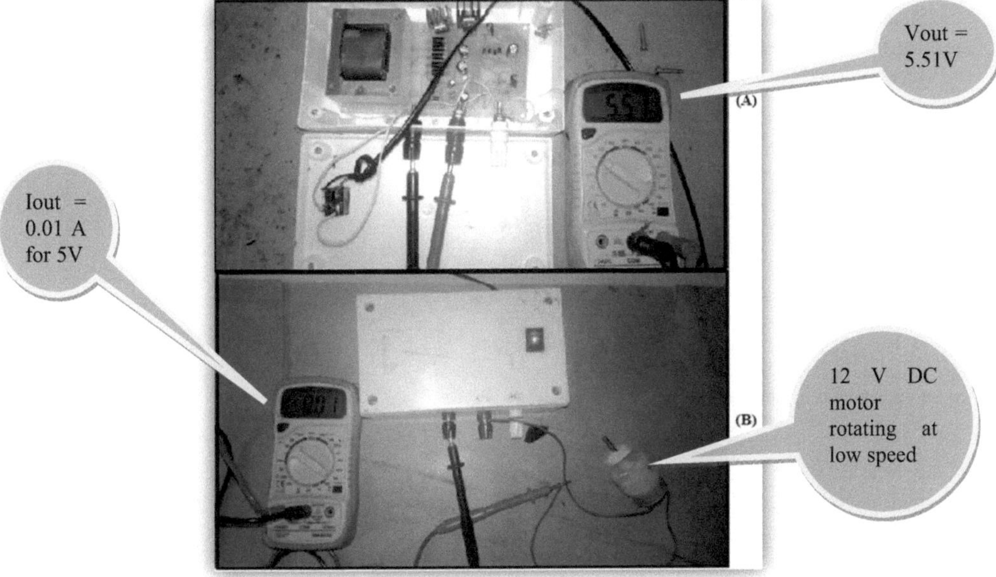

Figure 2. 5 5V output and current testing

The Figure 2.5 B, shows that a 12 Volts DC motor is interfaced to the 5 Volts output terminal of the power supply and an Ammeter (Multimeter) is connected in series to the DC motor to measure the current flowing through the DC motor (load) that is being displayed as 0.01A on the multimeter while the motor rotates at lower speeds.

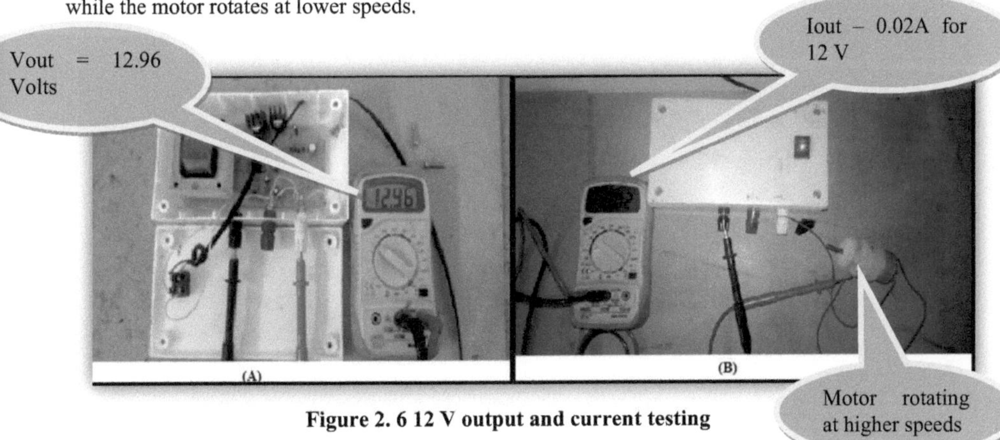

Figure 2. 6 12 V output and current testing

Once the testing of the current and voltage of the the 5 Volts output is completed, then the multimeter is interfaced with the 12 V output of the supply and the multimeter displays a result of 12.96 Volt (V) as shown in the Figure 2.6 A. The Figure 2.6 B, shows that an ammeter (multimeter) is connected in series to the DC motor which is interfaced to the 12 V output of the supply, thus the multimeter displays a higher value of current of 0.02A that is flowing through the DC motor, thus making the motor rotate much faster due to the supply voltage being much higher than 5 V i.e. 12Volts.

Output Test Results	Theriotical	Practical (Multimeter display)	Accuracy (%)
5Volts (V) output connector	5V	5.51V and current I = 10mA	+.5
12 Volts (V) output	12V	12.96V and current I = 20mA	+1

Table 2. 8 Test Results

The Table 2.9, shows the test results of the power supply in which the voltages and currents flowing across a load using a 12 Volts DC motor has been described along with the practical variation has been determined in terms of accuracy.

2.8 Conclusion

The datasheet with the product design specification and other details for the DC power supply is developed that uses a 230V, 6VA Transformer giving a dual output of 5V and 12 V using the Voltage regulator IC's L7805 and L7812. The schematic for the design is developed using Design Entry and the Design rule check (DRC) was performed with no errors, therefore further designing and development of the PCB board can be carried out using this schematic. There are generally four interfaces for this power supple i.e. three terminal connectors such as GND, 5V output and 12 V output, it has a plug that is connected to the 230 V AC mains while a switch is provide over the power supply to control the switching on and off of the power supply. All the components needed for developing this power supply are available in India, banglore which has been specified along with its details. The power supply's output current and voltages were tested by using a multimeter and a 12 V DC motor to test the current flow. For 5 V output, the tested output voltage was 5.51 V with an accuracy of +.5%, with a current flowing through the load of 10mA, while the test output voltage for 12 V output was 12.96 V, with an accuracy of +1%, with a current of 20mA flowing across the load from the 12 Volts output of the DC power supply.

3. Thermal Analysis and Vibration testing of the DC power supply

3.1 Thermal variation of critical components and spacing between the critical components

Once the product is developed it is necessary to perform thermal analysis on the product in order to determine if the product is meeting the deadlines providing a desired response and its reliability. The thermal analysis is performed on the product using ICEPACK V13 to determine the temperature variation of the critical components that has high heat or power dissipation.

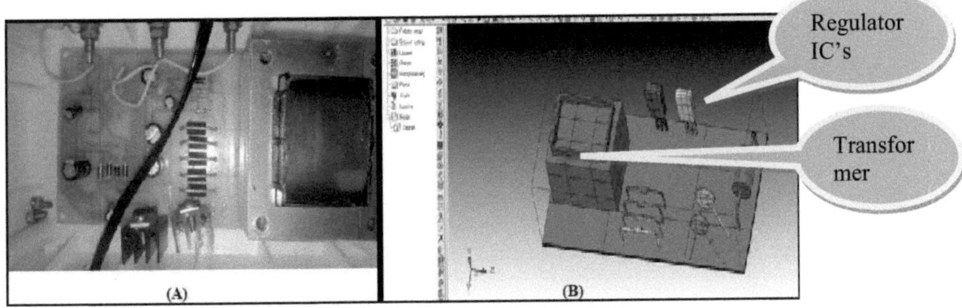

Figure 3. 1 Identification of critical components

The Figure 3.1 (A), shows the DC power supply with the thermal critical components in the system i.e. Voltage regulators IC's L7805 and L 781, 230V to 15V transformer. The lesser thermal critical components identified is the four diodes (H-Bridge), which has a low power dissipation, while the capacitors can be neglected for small system due to its negligible heat dissipation. The Components on the board are placed such that all the critical components such as the two regulators and the transformer is place at the end of the PCB for better natural cooling from the open top and boundary exposure for better convection to happen. The IC's are mounted on the heat sinks to increase the surface area and cooling mechanism due to the fins present in the heat sinks, the same goes with the transformer where the cupper windings is framed with aluminum to increase the surface area and to provide better cooling due to the fins present at the bottom of the transformer, since it is known that, heat flux is nothing but the energy dissipated per unit are and the surface area (A) is inversely proportional to the energy or heat flux dissipated i.e. if the Area is more then the stress is less and vice-versa. The DC power supply with some important components has been modeled using Icepack v13 as shown in the Figure 3.1 B, but the thermal analysis results is mainly examined for the critical components.

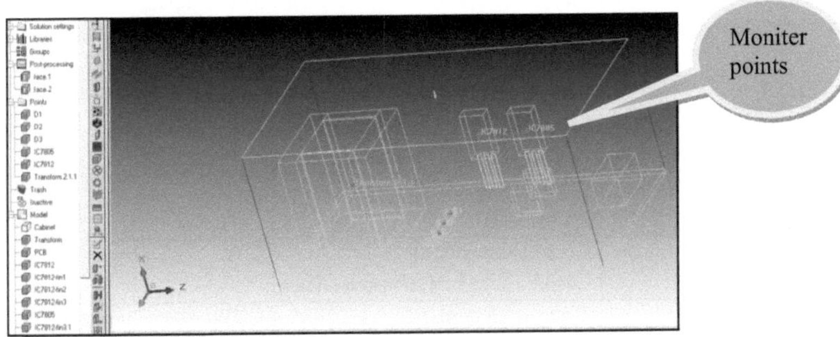

Figure 3. 2 Moniter Points

The Figure 3.2 shows the model of power supply in icepack with the components that have been assigned with monitor points in order to obtain the thermal variation curve of the monitored points. In the model, the components with the red dots indicate that those are the components that is to be monitored, whose thermal variation curves will be generated i.e. Transformer, IC's L7805 and L7182, Transformer and the H-bridge diodes.

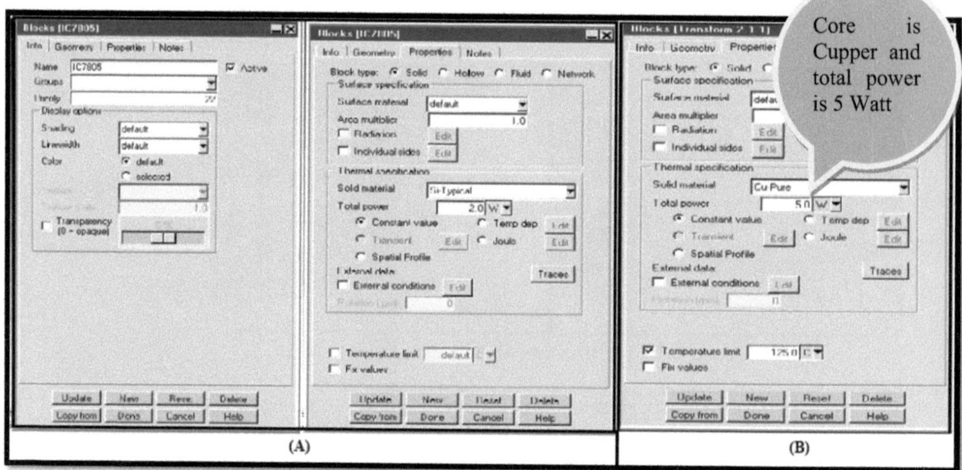

Figure 3. 3 Assigning material properties and total power

Before generating the thermal variation curves for the critical components in the power supply, it is necessary to specify the material properties and the total power parameters of all the components on board and Note that the thermal analysis variations on the critical components are performed for an ambient temperature of 24°C which is specified. The Figure 3.3 A, shows the properties blocks for the component IC L7805, in which the solid material specified is type Silicon (Si) Typical and the total power as 2 watt (W) and Figure 3.3 B, shows the material property assigned to the transformer core i.e. Cupper (Cu) pure with a total power of 5 W. The transformer has a power factor (PF) of 0.9 and a current of 5 Amperes (A), therefore total power is PF*I i.e. 0.9*5A = 4.5 Watt (W), but for worst case we consider the total power of transformer as 5W, since it's the most critical component that can generate high amount of heat. Similarly the material and the total power output for each component must be specified before running the solution for thermal variations of the critical components.

COMPONENTS	SOLID MATERIAL	TOTAL POWER Watt (W)
IC L7805	Silicon (Si)	2W
Leads	Aluminium	
IC L7812	Silicon (Si)	3.2W
Leads	Aluminium	
230V, 5A Transformer	Core : Cupper (Cu)	5 W
	Casing frame : Aluminium (Al)	
Capacitor (100nF)	Ceramic electrolytic	0.7W

Diode IN4007	Silicon (Typical)	0.05W
Capacitor (33,00uF)	Aluminised-Silica	0.8W

Table 3. 1 Material and total power (W)

The Table 3.1, shows the materials and the total power rating for each of the components present on board, which was calculated by identifying the total current and voltages of the components specified in the datasheets, while some components total power has been directly mentioned in the datasheets.

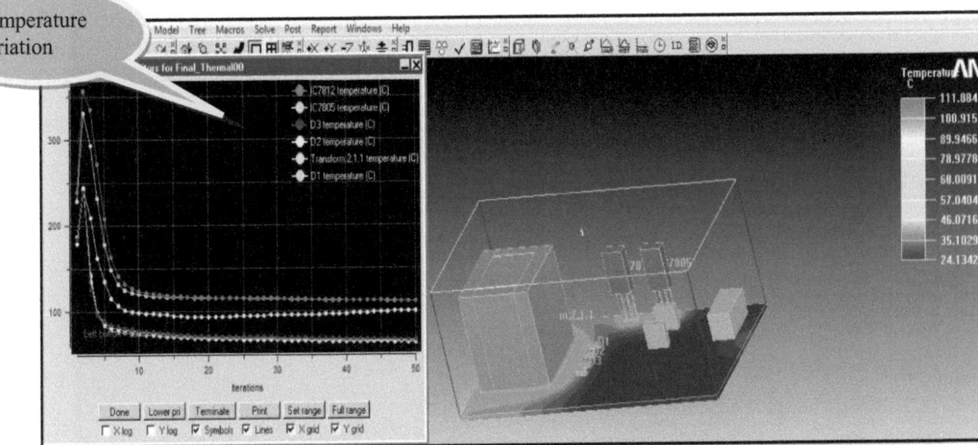

Figure 3. 4 Temperature variation of the critical components

Once the parameters such as the total power and the materials has been specified, the file is saved and the solution for the model is initiated by dropping down the solve icon in the software and then clicking "Run Solution" to obtain the thermal variation curves. The Figure 3.4 shows the Thermal variation curves for each critical component that has been assigned to monitor points.

Critical Components	Maximum Operating Temperature (^{o}C)	Software Simulation Temperature (^{o}C)
IC L7812	150	111.8
IC L7805	125	110.9
Transformer	130	100.9
Diodes (D1,D2 and D3)	175	68 to 78

Table 3. 2 Temperature simulation results

The Table 3.2, shows the thermal results of the critical components after the simulation and the maximum temperature below which each of these components can operate properly. In the table, it is seen that simulation temperature results are all normal and operating under the maximum limit temperature. In the Figure 3.4, the graph shows that, initially the temperatures of the components are at the peak or very high and as the simulation goes on, eventually the temperatures of the components decreases and becomes stable under the maximum temperature as it reaches the 20 iterations and remains stable beyond. In the thermal curves it is seen that the regulator IC's have the highest temperatures followed by the transformer and the Diodes.

3.2 Design of casing for power supply and its geometrical parameters

All power supplies generate waste heat which has to be dissipated. The heating effect becomes greater as more components are squeezed into smaller spaces. The result of miniaturization is higher levels of heat per cubic volume of space. The heat generated by components not only passes into the air around the components but is absorbed by adjacent parts, by the PCB and by the equipment case. As a result, various parts of the system end up operating at higher temperatures than originally anticipated, which adversely affects the reliability and service life. While some techniques can be used for better cooling of the power supply by creating openings for natural convection to take place

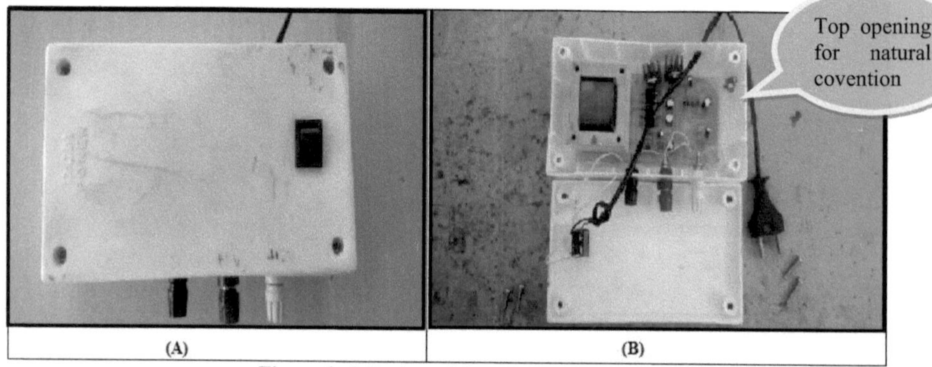

Figure 3. 5 Casing of the Power Supply

The Figure 3.5 shows the casing of the power supply. The power supply is designed for natural convection cooling, therefore it is necessary to keep in mind that while designing systems with natural convection cooling there must be free air present in abundant and the environment in which it is used must be an open space. The Figure 3.5 A, shows the power supply enclosed from all sides and Figure 3.5 B, shows the power supply with it top lid removed while creating an opening for natural convection to take place better.

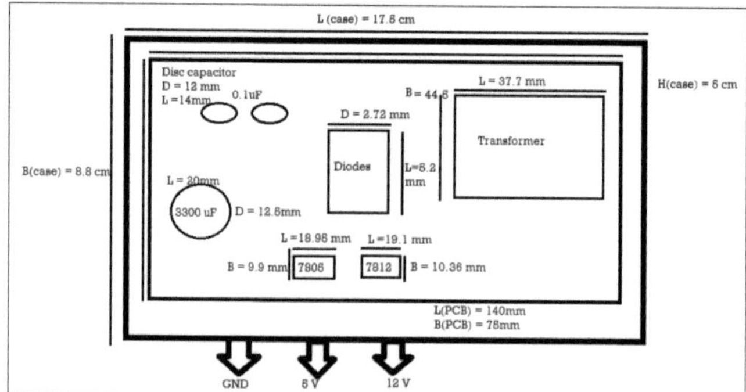

Figure 3. 6 Geometrical Dimensions of Power supply

The Figure 3.6, shows the geometrical parameters of the power supply taking from the outer casing of the supply, PCB board and the components present inside it. The block shown as diodes is the H-bridge that is situated ahead of the transformer, This figure 3.6 gives the idea and the parameters required for the modeling of the power supply in the software tool icepack. Length L(case) and Breadth B (case) indicates the dimensions for the outer case while L(PCB) and B(PCB) specifies the dimensions of the electronic board and followed by Diameter (D) for the capacitors and the diodes.

3.3 Design, Modeling and Simulation of mechanical casing for power supply using ICEPACK

Initially when the Electronic board is modeled along with the components present on it, a default enclosure (casing) or cabinet is present. In icepack the dimensions of the cabinet is generally 1mm greater than dimensions of the modeled board with components.

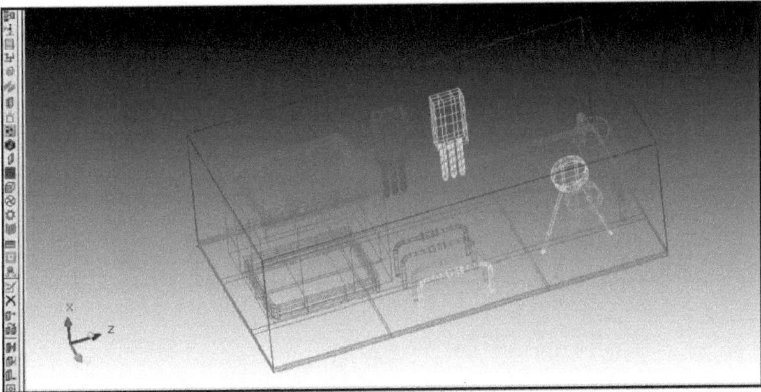

Figure 3. 7 Design of the Power supply casing without openings

The Figure 3.7, shows the design of the casing for the power supply in the wired form, in this design the casing does not have any openings, while all the sides are considered as walls. For conduction to take place, it is necessary to align all the part faces to face of the board. A good casing design is the one that must provide excellent cooling through air circulation for the on-board components.

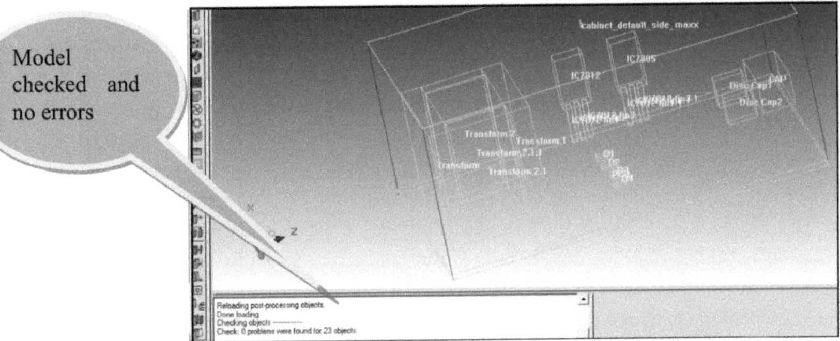

Figure 3. 8 Checking the Designed Model

Once the casing with the desired method of cooling and the desired onboard components has been modeled, the model design needs to be checked for errors, therefore in the software tool, in the menu Model→Check Model. The Figure 3.8, shows that the model consists of 23 objects and there were no problems found in the model. For better cooling mechanism, an opening is created on the top of the cabinet or casing for the model of the DC power supply.

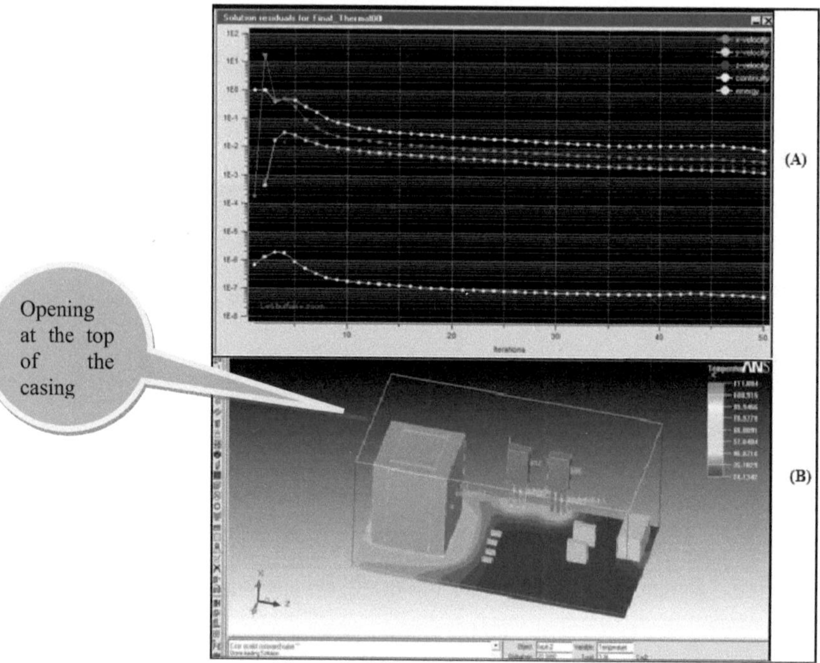

Figure 3. 9 Simulation of the casing with a opening

The Figure 3.9 B, shows the thermal variation of the different components in the power supply which as an opening at the top and walls at the sides. The opening at the top allows for greater volume of air to circulate through the components for better natural convection to take place. Once the openings are created then the solution must be run to check if the solution is getting loaded and to analyze the temperature variations with respect to this design of casing. The air flowing close to the surface of the board below the IC's is much warmer than that of the air flowing near the transformer followed by the diodes. The air flowing on the surface of the board beyond the diodes and the IC's are much cooler since the surface area is larger all the minimum heat generated by the capacitors to spread out easily due larger surface area maintaining a low temperature of about 24 degrees Celsius. The Figure 3.9 A, shows the simulation curves of the velocity at which the heat transfer or convection is taking place in different directions i.e. x, y and z directions and it also shows the momentum or continuity as well as the strength of the heat or energy. In the Figure B, it is seen that, since the opening is created at the top in the x-directions, therefore rate of convection is greater in x-direction initially as shown in the graph with a initial peak i.e. The large volume of air present leads to better convection. The energy or the strength of heat is much lesser due to natural convection and it is noticed that all initially all the curves started at a higher values eventually stabilizing and converging indicating that there is a controlled thermal transfer being taking place providing the appropriate response by the system with respect to cooling and better thermal heat transfer.

3.4 Optimized solution for mechanical casing design to meet the thermal variations

During natural convention (driven by buoyancy law of density), the molecules remain at minimum velocity and stable also known as free convention but when the velocity of warm air at the surface is increased by applying an external force e,g. by using fans is known as forced convention. Force convention is generally used in complex systems e.g. laptop fans for cooling the processor. Increasing velocity over a minimum constant velocity using fans will cause a shift from lamina to turbulent state that depends on the Reynolds number (Tells the status, whether lamina or turbulent).

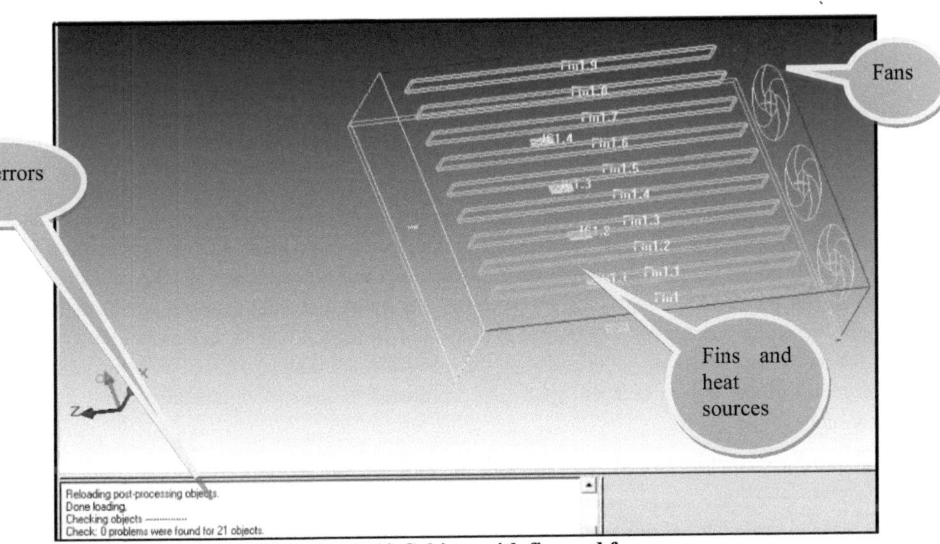

Figure 3. 10 Cabinet with fins and fans

The Figure 3.10, shows the model of a cabinet having five heat sources, ten fine and three fans. The fans are used for forced convention, while the fins are used to create fluctuations and increase the air velocity for better cooling. The fans exhaust the hot air while replacing it with cooler air. There is an opening created at opposite side of the fans to rush out the hot air. The model was developed based on a tutorial in software tool icepack. This method of cooling is used in complex systems or in systems having very high heating components with minimum surface area. However the modeled power supply can very well work by natural convention cooling, but adding an extra small fan to the system would change cooling from lamina or natural to turbulent i.e. forced convention, which will provide slightly better cooling performance.

Figure 3. 11 Power supply model with forced convention using a fan

The Figure 3.11, shows the power supply model along with a fan at one of its walls. The small fan rotating at lower speeds along with the opening provided at the top can provide the best cooling for the modeled power supply, however adding an extra fan can however increase the cost slightly. Note that, if the system components are performing at its best cooling temperature values using natural convention, then it not necessary to use any fans, because it will not make much difference. For the power supply, adding a fan on one of the sides as in the figure 3.11, will probably drop the temperature of the critical components by $2^{\circ}C$ compared to the model simulated using natural convention.

3.5 Experimental set up to carry out vibration analysis for the power supply

The component can get in to resonance condition when the natural frequency of the system is very close to operating frequency or close to excitation frequency and hence component can fail. To avoid such failures of the component, experimental techniques can be used to find out the natural frequencies of the component. The natural frequency (W_n) for a system is given by

$$W_n = \sqrt{k/m} \text{ rad/Sec}$$

Where, m = Mass (Kg or grms)

K = Restoring force (mN/mm)

The equation for an undamped system is given by $(-W_m + k)A\text{Sin}wt = 0$

Where A is the amplitude. If the amplide is zero, then it means theres no vibrations at all which is practically impossible, thus the system will be at rest or equilibrium only at the frequency W_n i.e. self equilibrium. If the system at self equilibrium is disturbed by an external force or impact, then the amplitude or energy at its natural frequency (W_n) increases and as time goes on, theres loss of amplitude bringing the system back to equilibrium after the impact.

Figure 3. 12 Power supply setup for vibration testing

The Figure 3.12 A, shows the power supply set up for the vibration testing. The PCB is mounted on the power supply plastic case which is flipped and the four corners of the PCB are bolted onto the case as shown in Figure 3.12 A. The Figure 3.12 B, shows the impact hammer i.e. an ordinary looking hammer with a sensor on its head, the hammer is connected to the Real time Vibrations Analyzer. The accelerometer is sensor mounted on the PCB to detect and measure the acceleration caused due to the impact by the hammer.

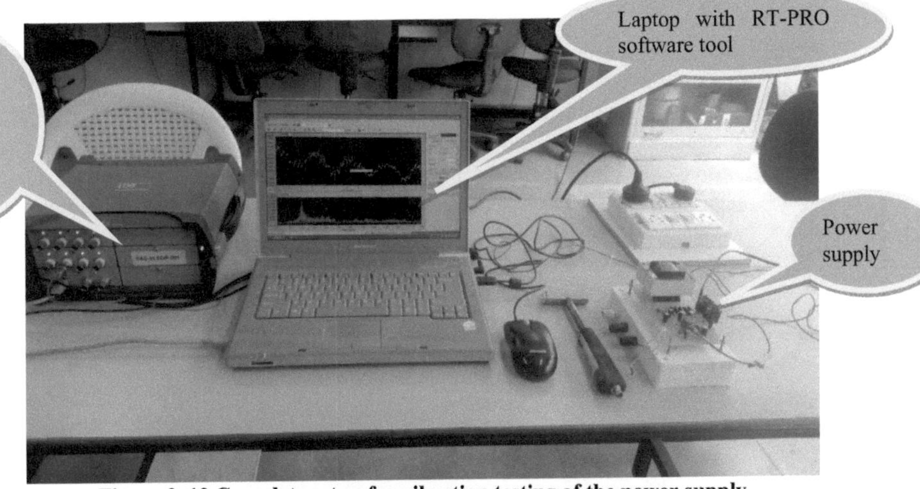

Figure 3. 13 Complete setup for vibration testing of the power supply

The Figure 3.13, shows the experimental setup for the vibration testing of the PCB and measuring the boards natural frequency. There's a laptop that has a software tool known as RT-PRO software, which is used for processing of the input analog signals from the impact hammer and the accelerometer. The accelerometer and hammer output is connected to the Real time analyzer and the laptop gets the analog inputs from the analyzer on which the RT-PRO software performs Fast Fourier Transforms (FFT) and converts it into frequency domain that is displayed in the software graphs for analysis. Vibration Analyzer processes and analyzes the signals received from accelerometer and impact hammer. Vibration analyzer used in this experiment has eight channels which means it can receive up to eight signals simultaneously. Four-channel vibration analyzer is quite common.

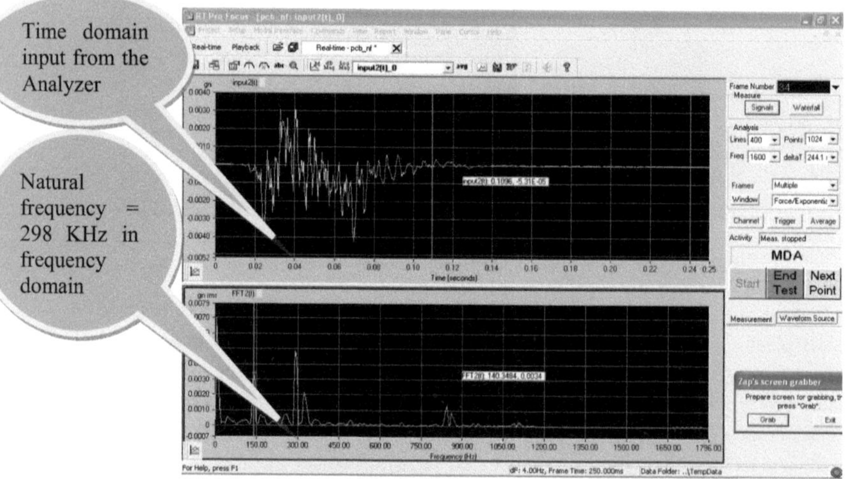

Figure 3. 14 Natural Frequency at equilibrium state of the system

The Figure 3.14, shows the time domain and the frequency domain inputs from the signal analyzer. The graph shows the simulation and the natural frequency during which the system or the PCB is at a state of equilibrium i.e. when no external impact is applied on the board. The Figure 3.14, shows that the natural frequency is 298 Hz, since the amplitude or energy at a frequency of 298 Hz is the maximum compared to its previous and the next harmonics. The one with the highest peak indicates maximum amplitude and its corresponding frequency indicates natural frequency of the component. Except first peak due to torsional vibration, there are eleven peaks. Therefore there are eleven modes. From figure3.14 it is obvious that maximum acceleration occurs at frequency of 298 Hz.

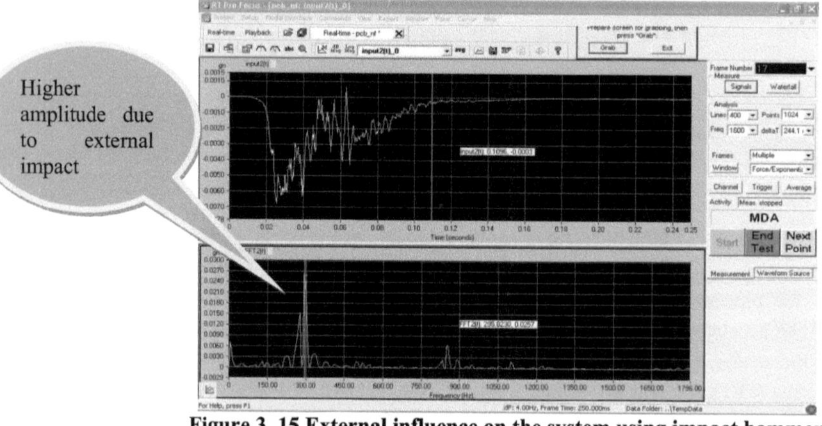

Figure 3. 15 External influence on the system using impact hammer

Once the natural frequency is identified, then vibration analysis based on the change in amplitude at the natural frequency (298 Hz) is analyzed by impacting on the PCB with the hammer at different forces that will lead to the change in the amplitudes at the natural frequency. The Figure 3.15, shows that the amplitude at the frequency 298 Hz has increased to due to an external impact on the PCB by the impact hammer. Note that the natural frequency detected using this approach may not be fully accurate due to the interference of the surrounding frequencies of the table, laptop and the casing over which the PCB is mounted, however the experiment was conducted to show the demonstration on how a vibration testing is carried out for integrated electronic systems and for better accuracy the surroundings and the place in which the vibration tests are performed must be proper e.g. the system being tested must be kept away from other high frequency equipments , since it may disturb the output results that will be obtained.

3.6 Conclusion

The temperature critical components in the power supply are identified and the 3D modeling of the power supply with its critical components is developed using the software tool ICEPACK v13. The thermal analysis is performed on the designed power supply model and the temperature variation curves of the critical components is obtained, where the IC L7812 has the maximum temperature of $111^{\circ}C$ and then comes the regulator IC L7805 with a temperature of $109^{\circ}C$, followed by the transformer at $100^{\circ}C$ and the diodes operating at a temperature range between $68^{\circ}C$ to $78^{\circ}C$, these simulations were performed for an ambient room temperature of $24^{\circ}C$. The designed power supply model critical components temperatures can be reduced further by providing better cooling system such as by using forced convention eg. Making use of a fan, where a model of the power supply along with a fan has been described. After the desired results were obtained from thermal analysis, the hardware DC power supply was exposed to vibration testing in which its natural frequency of vibrations was identified at 298 KHz, which was influenced by an external force using the impact hammer to verify the change in amplitude at that particular natural frequency.

3.6 Module Learning Outcomes

- Can apply research methods and techniques to perform need analysis and arrive at product design specifications of the product.
- Can describe the mechanical structures and its properties that needs to be integrated to an electronic subsystem to develop a complete electronic product or system.
- Can describe concepts with respect to different modes of heat transfer i.e. Conduction, convention and radiation.
- Can design, model and perform thermal analysis for the electronic systems using software tools such as ICEPACK v13
- Cna analyze and evaluate various opproaches used for interfaces in integrated electronic system.
- Can perform vibration testing of any electronic systems using a real time signal analyzer, accelerometers, impact hammer and a laptop with the RT-PRO software tool.

3.7 Summary

- ✓ In opto-electronic packaging, the principle used for packaging of receivers can be applied to the packaging of the laser drivers.
- ✓ The most mature technology presently being used for laser sources is the GaAs.
- ✓ One of the main challenges encountered during package is the materials lattice constant mismatch, for which the potential solution is by making use of alloys with nitride compounds.
- ✓ There are several tests that a product must carry out in industries among which for electronics, the EOS testing is very important to prevent thermal damage to the components and the and an industrial approach for calculating the reliability of the system is discussed.
- ✓ The critical components in the power supply is identified along with its material properties and power rating parameters and its dimensions using which it is modelled in the software tool ICEPACK v13
- ✓ The temperature variation curves for the critical components are generated during the simulation.
- ✓ An example and difference between natural convention and forced convention to the power supply system has been modelled and discussed.
- ✓ Practical Tests were conducted on the power supply whose results were analyzed and descripted.
- ✓ A complete experimental setup for carrying out vibration testing of an electronic system has been demonstrated and discussed in detail.

<u>REFERENCES</u>

1. Walter Willing, Jonathan Fleisher & Michael Cascio, *Electronic Part Failure Analysis Tools and Techniques"* Northrop Grumman Corporation, USA.

2. J. Salzman and I. Samid (*1996*), *GaAsN, A Novel Material for Optoelectronics on Silicon"* Israel Institute of Technology, Israel.

3. Craig A. Armiento, *"Silicon Wafer board: Challenges and Opportunities for Low-Cost*

4. *Optoelectronic Subsystems",* GTE Laboratories, Waltham.

5. Allan Armstrong, Scott Killmeye, Jeff Yee, *"Design Trends and Challenges for Parallel Optical Interconnect"* Vitesse Semiconductor, South San Francisco,

6. Mino.F.Dautartas, John Fisher, Hui Luo, Proyag Datta, and Arden Jeantilus (*2002),* *"Hybrid Optical Packaging, Challenges and Opportunities"*, Electronic Components and Technology Conference.

7. Mike Salib, Mike Morse, and Mario Paniccia, 1st IEEE International Conference on Group IV Photonics (2004), *"Opportunities and Integration Challenges for CMOS-compatible Silicon Photonic and Optoelectronic Devices"* Photonics Technology Lab, Intel Corporation, USA.

8. Wilton Workman , *"Failure Analysis Techniques"* Texas Instruments Inc, Texas.

9. William J. Vigrass (1997*)*, *"Calculation of Semiconductor Failure Rates"* Indianapolis,19-22 October.

10. Department Of Defense Test Method Standard Microcircuits (2004) ,*"Mil-Std-883f"* 31 December, Usa.

11. Rajeev Solomon, Peter Sandborn, And Michael Pecht (2000*)*, *"Electronic Part Life Cycle Concepts And Obsolescence Forecasting"* University Of Maryland, College Park, December, USA.

12. Jim Glancey (2006), *"Failure Analysis Methods What, Why And How"* USA.

13. Item Software, Inc (2007) ,*"Reliability Prediction Basics"* USA.